YOUR KNOWLEDGE HAS VALUE

Eva Weiss

Aus der Reihe: e-fellows.net stipendiaten-wissen

e-fellows.net (Hrsg.)

Band 131

Regulation of the iron transporter gene fepA is crucial for cell viability

GRIN Verlag

Bibliografische Information der Deutschen Nationalbibliothek:

Die Deutsche Bibliothek verzeichnet diese Publikation in der Deutschen National-
bibliografie; detaillierte bibliografische Daten sind im Internet über http://dnb.d-
nb.de/ abrufbar.

Imprint:

Copyright © 2010 GRIN Verlag GmbH
Druck und Bindung: Books on Demand GmbH, Norderstedt Germany
ISBN: 978-3-640-96530-4

This book at GRIN:

http://www.grin.com/en/e-book/175339/regulation-of-the-iron-transporter-gene-
fepa-is-crucial-for-cell-viability

GRIN - Your knowledge has value

Der GRIN Verlag publiziert seit 1998 wissenschaftliche Arbeiten von Studenten, Hochschullehrern und anderen Akademikern als eBook und gedrucktes Buch. Die Verlagswebsite www.grin.com ist die ideale Plattform zur Veröffentlichung von Hausarbeiten, Abschlussarbeiten, wissenschaftlichen Aufsätzen, Dissertationen und Fachbüchern.

Visit us on the internet:

http://www.grin.com/

http://www.facebook.com/grincom

http://www.twitter.com/grin_com

Regulation of the iron transporter gene *fepA* is crucial for cell viability

Eva Ursula Weiss

FepA is an *Escherichia coli* (E. coli) membrane protein which transports the iron-bound siderophore ferric enterobactin from the exterior into the periplasm. Thereby it helps to provide the cells with soluble iron. In the study, its gene, fepA, was ligated into pUC8. E. coli cells were then transformed with this construct and tested for presence and orientation of sense or antisense expression. The majority of colonies that have taken up pUC8 with a fepA fragment were found to contain the gene in antisense orientation. It was concluded that removal of the gene from its regulation leading to its over-expression results in unviable clones possibly due to membrane protein toxicity or iron toxicity.

INTRODUCTION

Iron uptake is crucial for the survival of most organisms (1). It is needed as cofactor in many important metabolic pathways including the electron transport chain (1). Availability of iron is complicated by the fact that at neutral pH the metal forms stable, insoluble complexes with hydride in water (1). In order to access the metal bacteria use siderophores, which when secreted compete with hydroxide to complex iron (1). These chelators are called enterobactin or ferric enterobacting (FeEnt) when bound to iron in *Escherichia coli* (*E. coli*) (2). Uptake into the cells is mediated by FepA in the presence of TonB (2, 3).

FepA is a 81 kDa protein located on the outer membrane of gram negative bacteria like *E. coli* (2). There it functions as receptor for FeEnt (2). It is very similar to other *E. coli* membrane proteins with respect to the distribution of charges and hydrophilic and hydrophobic parts on the molecule (2). Its structure is likely to also involve specific domains at the periplasm and exterior side in order to interact with its ligands and other proteins (4). It is thought to interact with TonB and ExbB both being required for function(4). More specifically, the interaction of FepA with the energy-supplying species, TonB, is thought to occur at the N-terminus of FepA (2). The amino-terminal globular domain forming a loop structure is believed to be important for the translocation of FeEnt whereas the loops of the β-barrel domain at the carboxyl terminus, to which the N-domain is associated, recognize the correct ligand (7, 8). The actual first binding of the ligand, on the other hand, was found to be performed by specific aromatic side chains on the protein (8). This causes a minor conformational change in the protein needed to move away the globular domain, which plugs the barrel, enabling FeEnt to be translocated (7). This

1

mechanism of closing and opening the channel makes the transport specific (3). Additionally, FepA is believed to interact with the other "fep" proteins FepB, FepC, FepD and FepE (4). The latter three are likely to form a complex at the inner membrane to allow FeEnt to proceed from the periplasm to the cytoplasm whereas FepB is found in the periplasm itself (4). The exact function of these proteins and how they interact with FepA are not known (2).

When bacterial cells are deprived of iron they boost the production of proteins involved in its uptake whereas the proteins' gene expression is repressed at high cellular iron(II) levels (2, 5). It is thought that a dyad sequence with a A-T doublet motif in the promoter is the site to which the transcriptional repressor Fur would bind to repress expression of the *fepA* gene when iron levels are high (2, 6). This regulation of *fepA* was found to be connected to other genes involved with iron by overlaps in promoter sequences (6). Interestingly, iron (II) itself binds to the repressor protein increasing its activity; thus, a high amount of iron(II) results in more repression of the genes involved in FeEnt uptake (5). Although so far only Fur is known to act as repressor in this process, it is possible that more regulatory proteins are involved (6).

For the study of *fepA* the gene was obtained from pPC104. The plasmid was first described by Coderre and Earhart. It was originally derived from *E. coli* genomic sequence. Caderre and Earhart have shown that pPC104 also contains the genes *entD fepA* and *fes,* in that order, together with the respective promoter. The products of all of these genes are essential for iron uptake in *E. coli* (9). *fepA* was moved to pUC8 in order to remove it from its endogenous promoter and

MATERIALS & METHODS

Amplification, isolation and digestion of pPC104

E. coli cells carrying pPC104, which contains the *fepA* gene, were incubated with chloramphenicol (60 μg/mL, overnight) allowing for replication of the plasmid but genomic DNA. This led to an amplification of pPC104 in the cells. It was subsequently isolated by centrifugation of the cells at 7,000 rpm with a Sorvall GSA rotor and then resuspended in buffer with lysozyme. The cells were lysed using the alkaline lysis approach (5 mg/mL lysozyme, 0.2 N NaOH, 1% SDS and later 2 N acetic acid). Precipitated genomic DNA was removed by centrifugation at 14,000 rpm in an Eppendorf microcentrifuge and filtration of the supernatant. Further purification was achieved by isopropanol precipitation, RNase (0.3 mg/mL) treatment and phenol/chloroform extraction with subsequent ethanol precipitation. The isolated and purified plasmid DNA was cut in a double digest with the restriction enzymes *Ssp*I (15 U) and *Eco 147*I (15 U) in Buffer Y/Tango (Fermentas Inc, Glen Burnie, MD) for 2 hrs at 37°C. The *fepA* fragment was isolated by agarose gel electrophoresis followed by gel extraction using the QIAEX II kit (QUIAGEN, Valencia, CA).

Preparation of pUC8 for cloning

Pure pUC8 was cut with the restriction enzyme *Hinc*II (10 U) in Buffer Y/Tango (Fermentas Inc, Glen Burnie, MD) for 2 hrs at 37°C. Since only one restriction site for this enzyme is present on pUC8, the digest led to an opening of the plasmid but no fragmentation resulting in a linear piece of pUC8 with blunt ends. To later prevent self-ligation of the plasmid, it was also treated with Shrimp alkaline phosphatase (SAP) for 1 hr at 37°C, which removed the 5' phosphate groups at the ends, in SAP buffer.

Preparation of competent JM83 E. coli

An aliquot of an overnight JM83 *E. coli* culture was grown to an OD550 of between 0.4 and 0.6. Then the cells were kept cold to prevent further growth. Competency was induced by pelleting the cells and resuspending them in TSS (1 % tryptone, 0.5 % yeast extract, 1 % NaCl, 10 % PEG, 30 % glycerol, 0.05 M MgSO$_4$, pH 6.5). The level of achieved competency was determined by heat-shock transforming 100 µL of competent cells with 15 ng, 1.5 ng and 0.15 ng of pure pUC8, adding 900 µL LB medium, plating 100 µL of each sample on an agar plate and counting the number of colonies after an overnight incubation.

Ligation of pUC8 and fepA fragment

Linearized pUC8 and isolated *fepA*-containing pPC104 fragment both had blunt and therefore compatible ends. They were ligated together using T4 ligase (1.5 U) in the respective ligase buffer (Fermentas Inc, Glen Burnie, MD). A two-molar excess of insert compared to vector were used (60 fmoles of *fepA* insert and 30 fmoles of pUC8) to increase the likelihood of a reaction between the two. The ligation was allowed to incubate overnight at 15°C.

Transformation of JM83 E.coli with the ligation product

The product of the ligation reaction was used to transform competent JM83 *E. coli* cells (2.6 * 10^6 transformants/µg pUC8). 100 µL of cell suspension were added to 20 µL of reaction mixture and after 35 min incubation on ice, the mixture was heatshocked at 42°C for 90 seconds and 900 µL LB medium were added and the dilution was allowed to incubate for 45 minutes at 37°C. Then 100 µL were plated on an agar plate containing X-gal (1.2 mg/mL) and ampicillin.

Determination of insert orientation

To determine the presence and orientation of the *fepA* fragment with respect to the promoter in pUC8 downstream of which it was inserted, six white, i.e. β-galactosidase deficient colonies, were grown overnight. The plasmids were isolated in a mini-prep using the boiling-lysozyme approach. Briefly, the cells were pelleted by centrifugation, resuspended in 25 μL buffer (8% sucrose, 50 mM EDTA, 50 mM Tris, pH 8.0) and lysozyme (20 mg/mL). After short incubation at room temperature 25 μL of the same buffer also containing 10% Triton X-100 were added and the solutions boiled for 45 seconds before 250 μL of another buffer (500 nM NaCl, 10 mM Tris, pH 8.0) were added. Chromosomal DNA was pelleted and removed and the plasmid DNA further purified by isopropanol precipitation and subsequent RNase (0.4 μg/mL) treatment. Purified plasmid DNA was run on an agarose electrophoresis gel and compared to unmanipulated pUC8 to test for presence of an insert in pUC8 by size. Orientation was assessed by digest in Buffer O with *Eco R*I (Fermentas Inc, Glen Burnie, MD), which has a restriction site in the multiple cloning site of pUC8 as well as asymmetrically inside the *fepA* sequence thereby producing specific banding patterns for each possible combination of pUC8 and insert.

RESULTS

The *E. coli* gene *fepA* encodes for the iron transporter protein FepA, which when expressed under normal conditions ensures the bacteria's survival by translocating FeEnt into the interior to provide the cell with essential iron. The gene is, together with its promoter and other neighboring iron-related genes, found on pPC104. To remove *fepA* from its endogenous promoter and to increase the number of copies, it was cut out of this plasmid and ligated into pUC8. The new construct was used to transform competent *E. coli* cells. Positive clones were identified by re-isolating the plasmids and running it on an agarose gel together with unmanipulated pUC8 control to which the samples were compared (Fig. 1).

In order to determine the nature and orientation of the inserts in each sample, a restriction enzyme digest was performed, which would produce characteristic fragmentation for each possible construct. These were subsequently run on a gel together with a DNA mass ladder for size determination (Fig. 2). The band pattern provided evidence that *fepA* was ligated into pUC8 in orientation of antisense expression. In further repeats, out of the 102 colonies that were found to have taken up a plasmid with pUC8 and *fepA* insert, 101 were in antisense and only one in sense orientation.

DISCUSSION

A DNA fragment containing the iron-transporter protein gene *fepA* was cut out of plasmid pPC104 and ligated into plasmid pUC8. The resulting constructs were used to transform *E. coli* cells. Analyzing the orientation for expression of the insert, it was found that a disproportional high relative number of *fepA* insert was present in antisense orientation compared to the presence of inserts in the positive orientation.

Assuming a random process, it would have been expected to find approximately equal numbers of pUC8-*fepA* constructs with insert in either orientation. The result obtained is consequently rather unlikely to be random; thus, it can be inferred that the process is in fact not random but that some sort of selection in favor of the antisense orientation takes place. This could in theory happen at different stages of cloning. Even though T4 ligase was previously found to be specific to a certain extent with respect to its substrates, it is unlikely to cause this phenomenon since only the orientation and not the ligation itself is affected (11). Transformation again is only affected by size and shape of the DNA that is introduced and neither is be affected by the orientation of insert. Consequently, the observation is most likely to represent a physiological phenomenon occurring inside the cells after uptake of the plasmid. This physiological phenomenon would lead to a decreased growth or viability of cells which have taken up a plasmid with the *fepA* gene in sense orientation.

Having the gene in sense orientation means that a sense mRNA can be transcribed and a functional gene product made, whereas this cannot happen when the gene is orientated antisense. FepA naturally occurs in *E. coli* and is necessary for the cells to take up FeEnt. Consequently, the mere presence of a functioning gene would not cause trouble for the cell. However, FepA is not made in normal amounts when *fepA* is expressed from pUC8 because firstly, this plasmid replicates more efficiently than pPC104, which means more copies of plasmid and gene than normally are present, and secondly, in pUC8 *fepA* is under regulation of a different promoter and is constantly expressed (12). One could argue that the gene is also present in sense orientation in pPC104 and nevertheless pPC104-containing cells grow. However, this plasmid is present at lower copy numbers and it contains the endogenous *fepA* promoter; thus, transcription repressor Fur, which is known to repress the expression of the gene at high iron levels, could bind to it and regulate the expression according to the needs of the cell (6). Consequently, despite the gene being present at a higher number than in a normal *E. coli* cell, the number is not too high and the endogenous promoter drives *fepA* expression allowing for normal gene regulation (9). Neither of these is the case in pUC8 resulting in over-expression of the gene, the likely cause of the observations.

Over-expressing a gene can have different negative consequences which would lead to decreased growth

5

or viability of the affected cell. FepA is a membrane protein and Wagner *et al.* have addressed different possible ways how, generally, over-expression of a membrane protein can lead to toxicity. Their data suggested that problems arise due to a limit in the capacity of the Sec translocon important for trafficking of membrane and secreted proteins. Overwhelming the translocon with over-expressed protein will lead to less being available for other, important cellular proteins. Negative consequences of this include a decrease in functioning of the respiratory chain and an accumulation of proteins in the cytoplasm which can form aggregates (13). Alternatively, the effects of *fepA* over-expression could be more specific meaning that the decrease in viability is related to FepA function. That it is necessary for transporting iron into the cell and that iron is needed as cofactor for many important cellular reactions (1). Having abnormally high levels of FepA would result in an increased and uncontrolled uptake of iron from the environment. This, however, can become problematic to the cell because the same reasons that make this metal such an important nutrient, a suitable redox potential, also make it toxic if not properly controlled. In redox reactions it can generate reactive oxygen species (hydroxyl radical, superoxide and hydrogen peroxide). These in turn can react with important biomolecules like proteins, lipids or even nucleic acids with effects ranging from non-functional enzymes to severe mutations and death (14). In cells with normal expression of FepA and therefore normal and resulted uptake of iron, the problem of iron toxicity is still present but can be controlled by other proteins in the cell, whereas their capacity is likely to be exhausted when several-hundred times more iron is present than under normal conditions. In order to find out if *E. coli* cells are less viable when *fepA* is over-expressed due to a general mechanism or due to the specific function of FepA, further experiments will have to be performed including the cloning of another membrane protein, which is not involved in iron transport and the cloning of other, soluble protein involved in iron transport into the cells into pUC8 to see if the same effect would be observed under any of these modified conditions.

Over-expression of the iron-transporter gene *fepA* in *E. coli* appears to have negative effects on the viability of the clones leading to clones with a non-functional gene on the introduced plasmid being positively selected for. The excessively made gene product FepA could, as membrane protein, exhaust the capacity of Sec translocons leading to a competitive inhibition of correct trafficking of other proteins, which results in less functional electron transport chain across the membrane or aggregation of accumulated proteins in the cytoplasm. Alternatively, its function to transport iron into the cell could lead to death by iron toxicity. Also a combination of both is possible. Even though the exact cause for the phenomenon still needs further investigation, the fact that reduced viability of cells over-expressing *fepA* is observable, reflects the importance of proper gene regulation.

REFERENCES

1. Postle K. Aerobic Regulation of the Escherichia coli tonB Gene by Changes in Iron Availability and the fur Locus. J Bacteriol. 1990;**172**: 2287-2293.

2. Lundrigan M D, Kadner R J. Nucleotide Sequence of the Gene for the Ferrienterochelin Receptor FepA in *Escherichia coli*. J Biol Chem. 1983;**261**: 10797-10801.

3. Payne M A, Igo J D, Cao Z, Foster S B, Newton S M C, Klebba P E. Biphasic Binding Kinetics beween FepA and Its Ligands. J Biol Chem. 1997;**272**: 21950-21955.

4. Ozenberger B A, Schrodt Nahlik M, McIntosh M A. Genetic Organization of Multiple fep Genes Encoding Ferric Enterobactin Transport Functions in Escherichia coli. J Bacteriol. 1987;**169**: 3638-3646.

5. Miethke M, Marahiel M A. Siderophore-Based Iron Acquisition and Pathogen Control. Microbiol Mol Biol Rev. 2007;**71**: 413-451.

6. Hunt M D, Pettis G S, McIntosh M A. Promoter and Operator Determinants for Fur-Mediated Iron Regulation in the Bidirectional fepA-fes Control Region of the Escherichia coli Enterobactin Gene System. J Bacteriol. 1994;**176**: 3944-3955.

7. Mia L, Kaserer W, Annamalai R, Scott D C, Jin B, Jiang X, Xiao X, Maymani H, Massis L M, Ferreira L C S, Newton S M C, Klebba P E. Evidence of Ball-and-chain Transport of Ferric Enterobactin through FepA. J Biol Chem. 2006;**282**: 397-406.

8. Annamalai R, Jin B, Cao Z, Newton S M C, Klebba P E. Recognition of Ferric Catecholates by FepA. J Bacteriol. 2004;**184**: 3578-3589.

9. Coderre P E, Earhart C F. Characterization of a plasmid carrying the *Escherichia coli* K-12 *entD, fepA, fes,* and *entF* genes. FEMS Microbiol Lett. 1984;**25**:111-116.

10. Lundrigan M D, Kadner R J. Nucleotide sequence of the gene for the ferrienterochelin receptor FepA in *Escherichia coli* . J Biol Chem. 1986;**261**: 10797-10801.

11. Harada K, Orgel L E. Unexpected substrate specificity of T4 DNA ligase revealed by in vitro selection. Nucl Acids Res. 1993;**21**: 2287-2291.

12. Vieira J, Messing J. The pUC plasmids, an M13mp7-derived system for insertion mutagenesis and sequencing with synthetic universal primers. Gene. 1982;**19**:259-68.

13. Wagner S, Baars L, Ytterberg A J, Klussmeier A, Wagner C S, Nord O, Nygren P-A, van Wijk K J, de Gier J-W. Consequences of Membrane Protein Overexpression in Escherichia coli. Mol Cell Proteomics. 2007;**6**: 1527-1550.

14. Papanikolaou G, Pantopoulos K. Iron metabolism and toxicity. Toxicol Appl Pharmacol. 2005;**202**: 199-211.

FIGURES

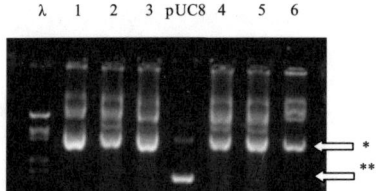

Figure 1: Determination of the presence of an insert in pUC8.
Isolated plasmid DNA from six different white colonies obtained on an agar plate after transformation with pUC8 and *fepA* insert ligation were run on an agarose gel electrophoresis gel (1, 2, 3, 4, 5 and 6) and compared to pure uncut pUC8 (pUC8) together with a λ/HinIII/EcoRI ladder (λ) as positive control. * indicates the band for the supercoiled form of the unknown plasmids and ** indicates the supercoiled form of pUC8. Additional bands for 1, 2, 3, 4, 5 and 6 are due to homologous recombination in the cells leading to multimerization of plasmid molecules. Only the lowest, monomer supercoiled form could be clearly identified and was used for comparison with the unmanipulated form of pUC8.

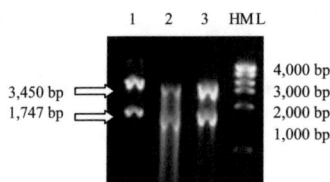

Figure 2: Determination of the nature and orientation of the insert in pUC8. The plasmid DNA samples isolated from transformed cells and positively tested for presence of an insert were cut with EcoRI to produce a banding pattern characteristic for a specific combination of pUC8 and another molecule of pUC8 or *fepA* insert and the different orientations. The results of the digests were run on an agarose gel (1, 2 and 3) together with a high mass ladder (HML) to determine the sizes of the obtained fragments. All of the samples produced the same kind of banding pattern with the lower band being between the 1,000 bp and 2,000 bp of the HML bands and the upper band in the range of the 3,000 bp and 4,000 bp ladder bands. The differences seen on the band represented differences in the amount loaded, which distorted the bands, rather than real differences in size.